Back Yard Wind Turbines

Harness wind power with simple and fun projects

Copyright © 2017 by Ahmed Ebeed

All rights reserved. This book or any portion thereof may not be reproduced or used in any manner whatsoever without the express written permission of the publisher except for the use of brief quotations in a book review or scholarly journal.

Table of Contents:

Preface

One   Old PC Fan to Wind Turbine

Two   TurbineOne: Wind Turbine anyone can make

Three   Failed PC Fan Blades

Four   5 Minutes Simple Wind Turbine for Everybody

Five   TurbineOne V2: The most straightforward wind turbine

Six   Wind Turbine Mobile USB Charger

Seven   Vertical Wind Turbine from Big PET Bottle

Eight   Half Brainer Wind Turbine

Nine   Failed - Mini Wind Turbine

Preface

Wind power has always been an inspiration for people all around the world and a unique symbol for freedom.

I've always wanted to make something that could harness this power even to the slightest amount of energy.

My focus has always been around reuse and recycling and I felt there were abundant resources waiting for mining.

2013 was the start for me when I made my first project. It has been my passion since then. And then I didn't look back.

Today I want to share with you those simple yet fun projects and encourage you to try with your kids and students.

All I want to say to you is: "Find your Passion" and "Follow it"

Do not give up your dream even if you had to work in other area or profession for living.

Keep working and searching for ways that you can practice your passion until you get chance to make significant benefit and then real profit from it.

If you kept learning, working on your dream and helping people with this dream you will get rewarded at the end.

Thank you for reading this book.

Ahmed Ebeed

Old PC Fan ----> Wind Turbine in 10 Minutes

I looked at some old PC Fans I have and thought that they can be used as Small Wind Turbines.

It has been my dream for a long time to make a wind turbine generator even to light an LED.

The PC Fan is Brushless DC (BLDC) Motor. It can be converted to a generator in 5 Minutes.

Stuff you need:

PC fan: old or new

Crocodile Clips

Soldering Iron

Step 1: Concept

The concept is simple. You can skip this part and start directly with the conversion. The BLDC motor used here has a stator winding and a Permanent Magnet Rotor. The motor is supplied by 12V DC. But the magnetic field rotation is generated by electronics (Electronic Commutator). As the name implies, the electronics components convert DC into AC which makes the magnetic filed in the stator rotate.

The electronic commutation is achieved by a small IC.

To get the induced current from the motor used as a generator , you must remove this IC.

Step 2: Disassemble

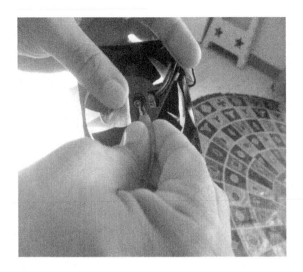

This is what I'll show you how.

First remove sticker on the back of the fan.

Then you'll find a small piece of plastic lock that holds Fan shaft secured, don't break it. Remove it with a crocodile clip.

This crocodile clip is very useful in this job. Many people have been asking how would they remove this plastic lock and I've answered them that they can easily use the crocodile clip to do it without breaking it.

Step 3: Winding Soldering

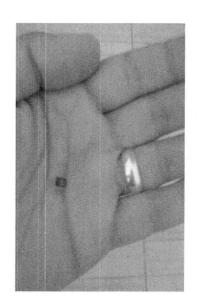

You can see 4 poles of winding connected in series and have only 2 terminals. To get the induced current, connect supply wires to those terminals and let the fan rotate.

With a solder iron, gently remove solder under IC pins and then remove the IC.

Remove any surface mount resistors or transistors.

Remove the supply wires form there place to put them in the holes of the removed IC.

Connect the supply wires in the winding terminals.

Make sure that you connect the terminals (from which you will get the generated Voltage) in a way that makes the two sets of winding connected in series.

Step 4: Final

Assemble the fan in its place and lock it with the piece of plastic I told you about.

Put back the sticker.

Connect LED terminals to the Supply wires. Don't worry about +ve and -ve terminals, the LED will light if you connect it any way, trust me.

Roll the Fan

I connected a voltmeter to see the output voltage. It was about 3 Volts on Average.

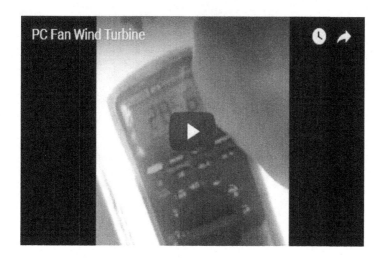

Thank you for reading.

TurbineOne - Basic Wind Turbine That Anyone Can Make

This is my first working practical <u>wind turbine</u>. I really love green projects and <u>renewable energy</u> stuff. Last year I've made a small modification on an old PC fan to convert it into a small wind turbine. It had enough output power to light an LED. It was a huge project for me at that time because I've always wanted so much to get even little power from wind.

The huge number of people on instructables who successfully built different sizes and shapes of practically working wind turbines has motivated me to build my own next level wind turbine to have higher scale of power output.

That's where TurbineOne came from.

TurbineOne is my first practical power generating wind turbine DE.

I named it TurbineOne because I intend to build many other turbines.

I'll explain how I built it in the next steps.

I know when it comes to technical appeal, engineering calculations or technology practices TurbineOne is not very awesome.

Believe me; I'm not so handy when it comes to mechanics and using power tools.

Please make good comments and productive criticism.

Step 1: Generator

This is the most important piece of equipment for your wind turbine DE.

Actually, it was the first thing I started to look for when I decided to build my own wind turbine.

I thought to buy a DC motor from any hardware store who sells this kind of motors as a spare part for any appliance and I thought to get them from eBay

(e.g. dishwasher , blender ... etc) .

If you couldn't find an old motor, you still have the option to buy a new one.

Then I found an old blender motor that has a permanent magnet inside it.

The motor generates electricity when it is turned by hand.

I measured the output and found to be nearly 14 Volts on the Voltmeter.

Step 2: Material

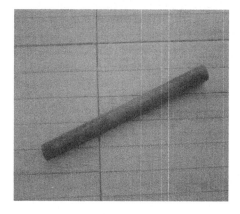

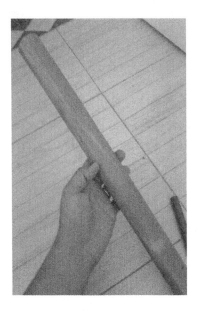 

This wind turbine is 100% recycled. I got all parts from scrap and used stuff.

It took me a long time to collect some of the materials used for building it, but you can just buy them or be lucky to find them easier than I did.

PVC pipe ---------> I found an old PVC pipe of suitable length to be used as turbine blades.

5 CD ROMs -------> I used old CD ROMs and DVD as wind turbine hub.
I found out that CDs is thicker than DVDs.

Fax paper plastic roll -----> used as a coupler between CD ROMs and motor shaft.

Some screws.
Some wires.
Old metallic rod used as a tower
Plastic tie raps

I added this instructable to the Leftovers contest because it is made of 100 % up-cycled and old leftover stuff.

Step 3: Tools

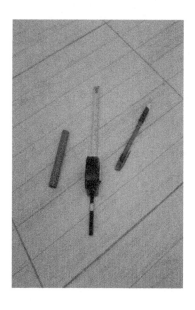

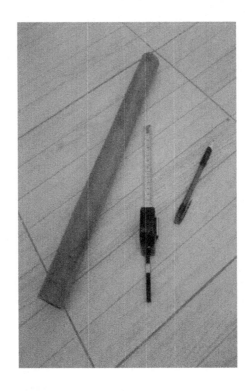

This project is made using fairly power tools. Please be careful when using this stuff.

- Saw

- Screw Driver

- Sand paper

- Pliers

Step 4: Hub Assembly

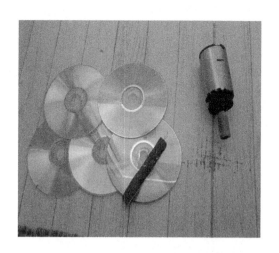

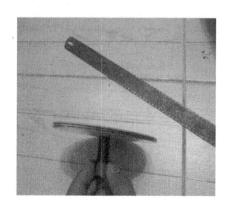

I started by the turbine hub.

Cut the plastic fax tube to 5 cm long.

I put the plastic tube around the motor rotating shaft.

Use the sand paper in CD ROM center to make the plastic tube fit into it.

Put CD ROMs and DVDs around the plastic pipe and motor shaft.

Step 5: Blades Assembly

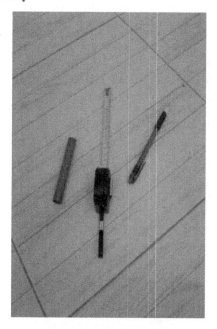

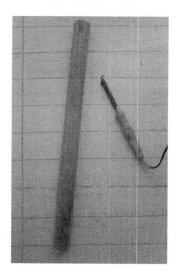

I wanted to cut the turbine blades into the usual turbine blades shape.

I really liked the idea of using PVC pipe as a fan blade. I got this idea from the internet.

But when I got the old PVC pipe I stated by drawing the fan blade on a template to draw it on the PVC pipe.

Then I couldn't get the tools to cut pipe in the fine shape of the fan blade.

So I've chosen to make the easy way and cut the PVC pipe into straight three equal pieces using hot soldering iron.

But how are these pieces going to generate rotational motion from the wind.
I decided to install each blade on the hub so it becomes nearly perpendicular to the hub and the round shape of the pipe does the rest.

Step 6: Turbine Assembly

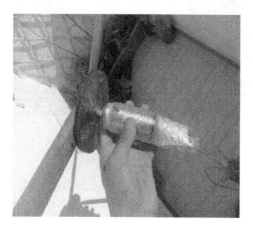

I installed the hub around the motor and secured them together using the fax paper plastic tube.

Then I cut the extra piece of plastic from the small plastic tube.

I drilled the metallic tube to install screws to fix the two tubes together.

I installed the turbine with its pole on the roof of my apartment.

It really rotates when the wind is fairly blowing.

TO DO:

I'll make a battery charging circuit and connect a sealed Lead-Acid battery to make a steady supply power source.

The ready made wind turbine kit will always be a good option for me.

Thank you for reading.

Failed: PC Fan Wind Turbine Blades!!!

I really want to share those new projects I make even they were not so unique but sharing what I make keeps me feeling alive.

So Let's get started.

As you all know, some projects work and some just fail. That's how life is all about. However, in instructables, I used to post only my working projects. That is how I turn my projects into instructables.

But I thought that a failed project is not actually a total failure. That project I made and sure learned something from it. So it must make a good instructable for all of us to learn from.

So here is my first **failed** instructable.

I previously made a simple modification to an old PC fan that turned it into a *fast to build small wind turbine*.

The project worked perfect , I wanted to make a better mechanical improvement to the turbine by adding blades to the rotor instead of the original blades installed on the PC Fan to increase fan speed and then the output voltage.

Tools:

Pliers

File tool

Scissors

Step 1: Modify the PC Fan

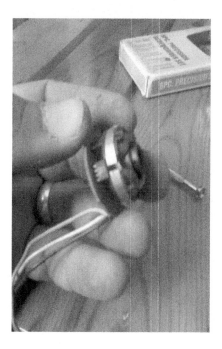

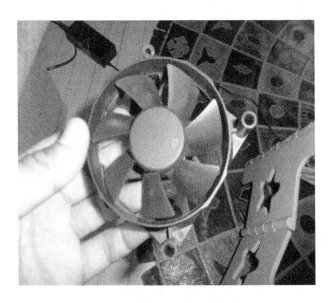

This step is described in detail in another chapter.

However, its main purpose is to turn the PC Fan into a small wind turbine.

This is done in 2 main steps.

Step 1:

Opening the PC Fan without breaking it and then removing the IC (Electronic Commutator) that converts DC into Square wave voltage to make a rotating magnetic field into the PC Fan stator side.

Step 2:

Connecting the stator winding together to sum up the generated voltage and get it out of the PC Fan. Then closing the Fan again.

Step 2: Remove the Fan Fins and Case

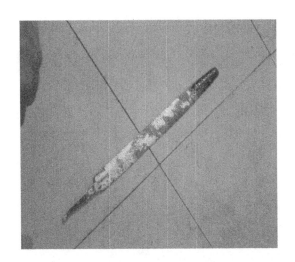

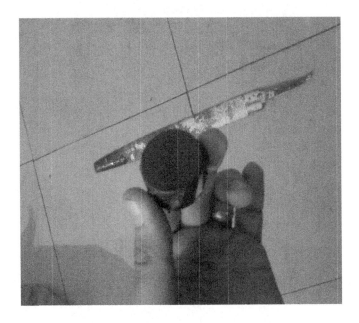

You need to remove the fan fins and case.

Tools:

Pliers

File tool

Here, I removed the fan case and fins with the pliers and left only the inner rotor connected to the stator and it PCB.

Then I used the file tool to soften the surface of the motor hub.

Step 3: Prepare the Large Turbine Blades and Connect Them to the PC Fan

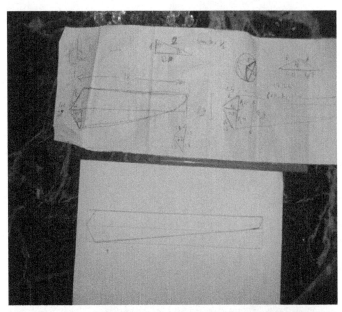

I made the large plastic wind turbine fins by drawing a sketch of the fins and cut a piece of plastic bottle to the sketch using scissors.

Step 4: Assembling, Testing and Facing the Truth

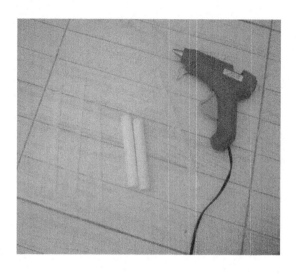

I connected the three wind turbine blades to the PC Fan hub using **glue gun**.

Then used an old fax paper roll and glued it the stator PCB to act as the shaft.

So far so good.

The only problem aroused when I put the fan into action. When I put the fan in face of the wind, the blades connected to the hub using the glue couldn't withstand even the normal breeze of wind. The three turbine blades were ripped off the PC Fan rotor hub.

I couldn't figure out a new way of connecting the **turbine** blades to the PC Fan rotor hub.

Then I declared a **failed** project.

Anyway, it was very educational.

5 Minutes Simple Wind Turbine for Everybody

I am fascinated with wind turbines and green energy. Today, I tried a new simple project that anybody can make.

I found this old fan blade assembly and decided to use it in something useful.

Step 1: Tools and Components

Here are all you need

Components:

Generator

Old fan blades

LED from

Tools:

Voltmeter

Screwdriver

Step 2: Assemble Your Turbine

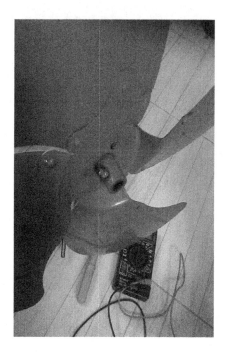

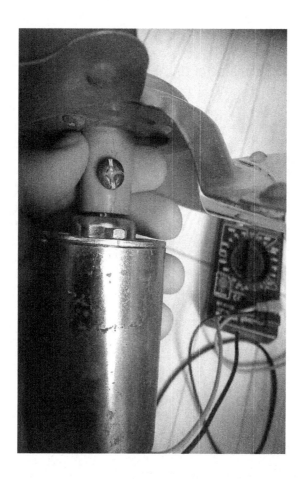

Now connect the fan blades to the motor shaft and use the screwdriver to install tighten the screw.

Now you've got your fully functioning wind turbine.

Let's test it in the wind.

Step 3: Test the Turbine in the Wind

In this step, I tested the **wind turbine** with a **Voltmeter** and an LED.

As you can see when the wind blown so strong, the LED was brighter. Also, the Voltmeter has measured about 30volts open circuit (with no load).

TurbineOne V2: Super Simple Wind Turbine You Can Make Now

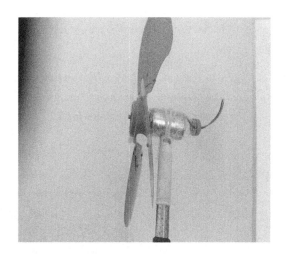

Last year I made **TurbineOne**, a completely recycled material <u>wind turbine</u>. Actually it was so simple and easy to make. But it was also so fragile due to the fact that its hub was made of compact discs **CDs**. As you can see in the photo, strong wind in one day has cracked the hub leaving the blades unconnected. The project was very inspiring and motivational. I've learned new stuff.

This year I wanted to make something more durable. With **Simplicity** and **Recycling** in mind I came up with a new idea of making a wind turbine using ready made blades from an old fan. Here comes the concept idea behind **TurbineOne V2.**

Step 1: Concept

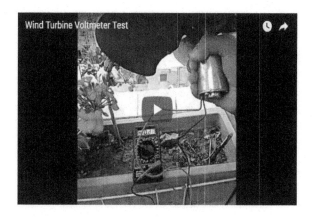

I've tested making this turbine setup in one of my previous instructables. I wanted to make a quick test for the concept.

Of course, it was something obvious and very straight forward but I wanted to really make sure that the wind is able to rotate the fan blades and resist the high motor torque.

I was in doubt because the motor had high rotation torque resistance.

It worked excellent even in slow wind. I was really surprised and it made me more motivated to finish this wind turbine setup.

Step 2: Components

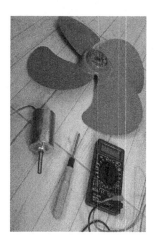

Here are the components for this project, you can see that nearly all of them are recycled materials :

Motor

This was the only new component I bought for this project.

Old steel groom --- This is the main pole for the wind turbine

Tie

Old CD-Rom player metal cover --- This is the directing rudder. For wind turbine automatic direction through all wind situations.

PET bottle ---- To cover and protect the motor against dust and water.

Some Wires

Step 3: Construction

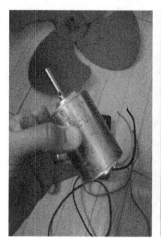

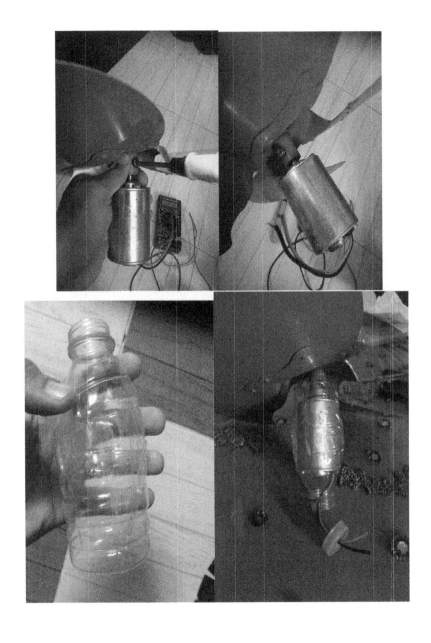

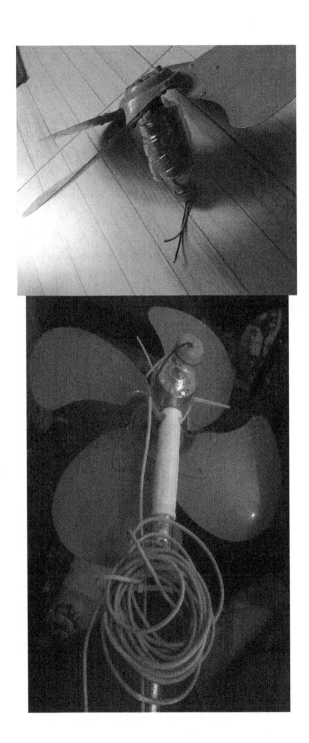

Very simple construction.

First, make the cover of the motor using the PET bottle.

Cut the bottle in two halves and cover the motor with them. Use a solder iron to shrink the bottle around the motor.

Step 4: Install the Rudder

I wanted to make something new in this version so I've decided to install a directional rudder that can direct the wind turbine automatically.

In the previous version there were no rudder installed, so I had to adjust its direction everyday according to the wind direction.

I used an old CD drive cover as a rudder and connected it to the wind turbine using tie raps.

Installed the generator to the pole using tie raps around it.

Connect a long wire to the motor terminals.

Step 5: Test

Put the **wind turbine** pole on a high place where you can have enough wind to run your wind turbine.

Wind Turbine Mobile USB Charger

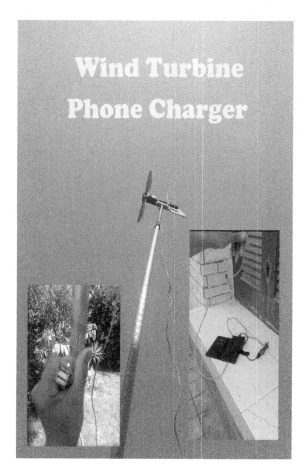

I love wind turbines. Last month, I built a new working wind turbine from recycled material. Today I've added some small gadget that made it a practical Mobile USB Charger powered by wind.

Step 1: Collect Stuff for Your Charger

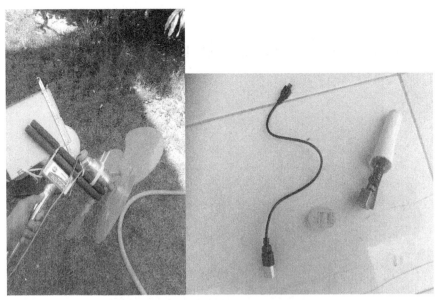

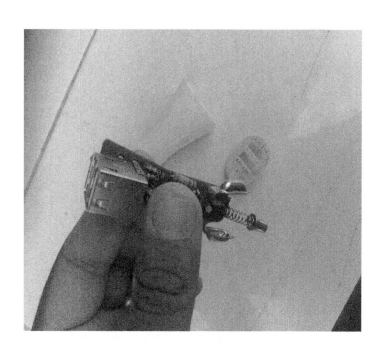
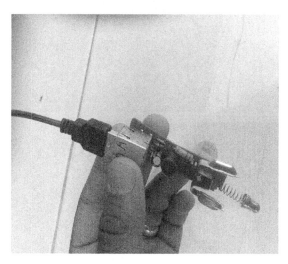

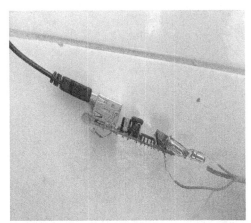
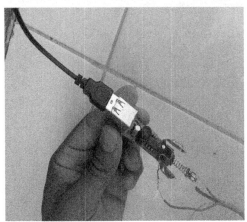

As usual, I've collect this instructable material from recycled stuff and other projects I've completed.

You can find that stuff around at your disposal or you can buy them.

For the wind turbine, I've added detailed instructions of how I made it in this chapter.

Material:

- I got this wind turbine I've built last month.
- USB Mobile charger for car
- Some weights
- Some wires
- Long metal pipe

Step 2: Connect and Test

Connect:

Connect the output of the wind turbine to the input of the USB charger terminals.

That's it. You've now got a working wind turbine Mobile Phone Charger.

Test:

Roll the wind turbine fan to test the system for functionality. You can see the Mobile Charger LED is illuminating as you roll the fan.

Congratulations. It works!!!

Uplift:

Install the wind turbine on the long metal pipe so it can get higher wind speed at the higher altitude. Install the metal pipe and secure it with heavy weights.

Step 3: Charger in Action - Running in Actual Wind

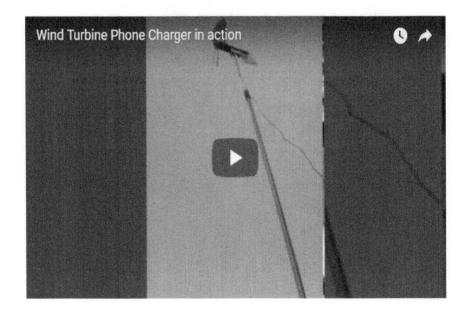

Put your wind turbine at a windy location and connect the Mobile Phone with the USB cable to the charger. Wait for the wind to blow.

As you can see in the video, at a lower wind speed the wind turbine rolls at a fairly low speed. But the charger is working pretty well.

Through the practical experiment, connecting the charger to the wind turbine chargers the phone faster than when connect it to a small solar panel as I did before.

Vertical Wind Turbine from Big PET Bottle

I love wind turbines. And I love making them specially from recycled materials. I just can't help myself when I get that feeling to reuse old stuff to make useful projects.

I have great interest in renewable energy. I've tried making some wind turbines before and today I'm trying to make a VAWT (Vertical Axis Wind Turbine). I'll show you my trials and failures until it worked.

Step 1: Components and Tools

Components:

Here are the components for this project, you can see that nearly all of them are recycled materials :

Motor

This was the only new component I bought for this project.

Old steel groom --- This is the main pole for the wind turbine

Tie rap

Old CD-Rom player metal cover --- This is the directing rudder. For wind turbine automatic direction through all wind situations.

PET bottle ---- To cover and protect the motor against dust and water.

Some Wires

1 mm Copper wire

5 Gallon water Bottle: I used PET Bottles that can be used only once. So they can be recycled for other purpose but not reused for drinking water.

Tools :

Pliers

Scissors

Gardening Scissors

Step 2: First Trial – Failed

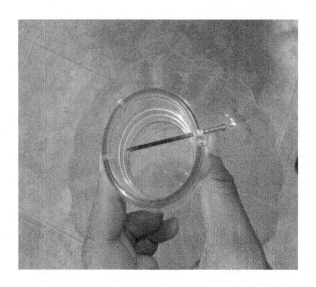

I used the Gardening Scissors to cut the PET bottle and made vertical slices so it may cause it to rotate if hanged freely into the air flow.

I used a hot nail to make two holes in the bottles neck so it can be attached to the generator's shaft.

After trying the turbine in the air, it didn't move.

I guess we can call it a failure, right?

Wrong. It's just another way of making Vertical Wind Turbine that doesn't spin.

Keep Reading

Step 3: Second and Third - Another Failure

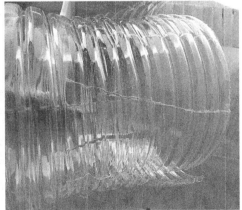

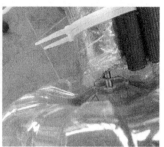

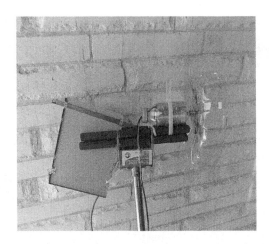

Then I tried another way of cutting the PET bottle.

I made also some vertical cuts while getting them out to make some fins that may cause it to spin.

No rotation either.

The third thing I've tried is to cut the PET bottle in some sloping lines. You see, I have many bottles of these.

I even tried another thing that might not be considered as a wind turbine. But the more I feel like I come closer to my target, the more I feel motivated.

I've tried to cut the bottom of the bottle and make a rotating disk that connects to the generator and can be rotated by hand.

This one really works.

Step 4: Forth Trial -- Successful!!!!

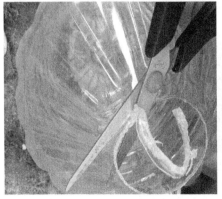

"Failure is no a failure if you learn from it"

This is just the quote I keep remembering when I start to build one of my projects.

Every project I make - fail or success - I learn a lot from it.

This is the final step where I could do the VAWT Vertical Axis Wind Turbine.

I cut the PET bottle into two halves.

Removed the bottom plate.

Connected the two halves together back-to-back using 1 mm Copper wire.

This made the rotor's fins.

I connected the fins to the generator using 1 mm Copper wire.

This one has worked.

Half Brainer Wind Turbine

If I managed to build it and it works so you can make it too. This is no joke. I've built this one with only half brain.

I came back from my exhausting night shift at work and wanted to spend some time in something useful and guess what I could come up with. Yes. A working wind turbine.

I really come from night shift feeling that I am only thinking and seeing the world with only half a brain. Let's get started.

Note:

This is how I look after coming home from a night shift.

Step 1: Collect Stuff

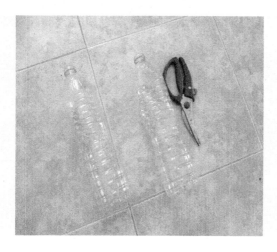

Components:

Here are the components for this project; you can see that nearly all of them are recycled materials:

Motor

This was the only new component I bought for this project.

Old steel groom --- This is the main pole for the wind turbine

Tie rap

Old CD-Rom player metal cover --- This is the directing rudder. For wind turbine automatic direction through all wind situations.

PET bottle ---- To cover and protect the motor against dust and water.

Some Wires

1 mm Copper wire

1.5 Pet Bottles

Tools :

Pliers

Scissors

Gardening Scissors

Step 2: First Trial

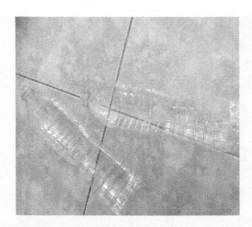

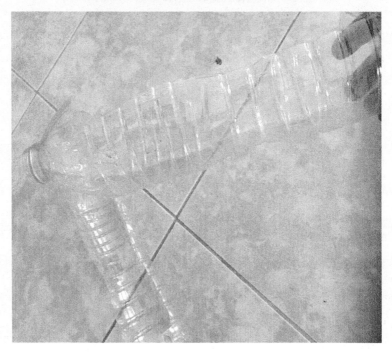

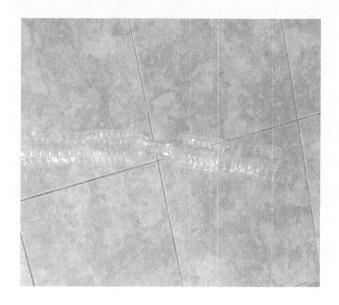

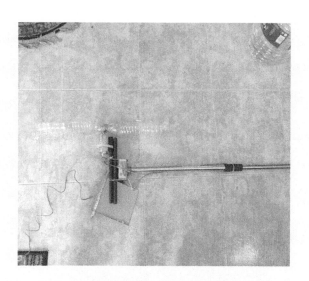

Actually I intended to build a vertical axis wind turbine using two PET bottles.

First, I cut the two bottles to make fins.

Then I welded the two bottles from their top using heat.

And made holes for generator and for Copper wires.

Then installed the generator and fixed it using Copper wires.

Failure. Just another none rotating vertical axis wind turbine.

The bottles didn't make enough torque to rotate the generator.

Step 3: Second Trial - Success

Using half brain, I've decided to cut the bottles bottoms hoping to make some rotation.

Using trial and error, I've discovered what works and what doesn't.

And then I cut the bottles to shape each one if them into something that looks similar to wind **turbines** blades to generate enough torque for rotation.

Success. It works and generates electricity.

Also another thing I discovered about this wind turbine is that it is completely safe to you and to your kids.

The blades are very elastic and delicate. Even if it hits something while rotating there is no problem at all.

Failed - Mini Wind Turbine

I love building small wind turbines for educational and experimental purposes. Some of them work and some just don't.

I learn from both those work and from those don't work.

So I actually make use of all of them.

In this instructable I'll show you a small wind turbine that I've recently made but didn't produce enough voltage to light an LED even though the wind speed was fairly enough to make other turbine work better.

Step 1: Components

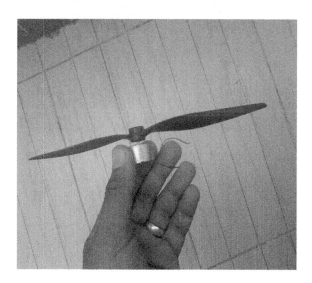

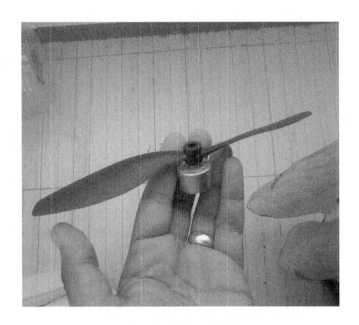

Motor Small motor from an old CD-ROM drive LED (I used white and then I used red)

Quadcopter blades 8045L 8 x 4.5L Metal Pipe

PET bottle

Thread

Step 2: Connect Parts Together

Cut the PET bottle to make a small piece from the top of the bottle. This piece will be used to fix the motor on the metal pipe.

Insert the motor inside the bottle top as shown in the image.

Connect the LED to the motor terminals.

Using the thread, fix the piece of PET bottle to the metal pipe.

Put the metal pipe in a clear open area where there is a high air flow rate.

Step 3: Testing and Failure

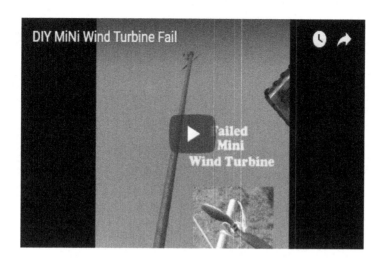

Testing the Mini Wind Turbine in the wind. I thought the turbine speed was enough to light the LED. I was wrong.

Although I've tested the motor to light the LED by turning it with hand, it really worked. But when I connected it to the blades, its speed was dramatically reduced.

The blades were heavy for the motor and they rotated at relatively low speed which caused the motor to output very small voltage.

This small voltage wasn't enough to light a white LED or a red LED.

I've learned from this one that not all rotors are compatible to all motors.

And that not all blades can be used as blades for **wind turbines**.

Those quadcopter blades were heavy and slowly driven the motor to low rotational speed.

But I am on my quest for making better and easier **wind turbines** for all people who are willing to make.

Made in the USA
Monee, IL
18 December 2020